STARS AND PLANETS

Written by Kris Hirschmann

Designed and Illustrated by Daniel Jankowski

tangerine Press
an imprint of
■SCHOLASTIC
scholastic.com

All rights reserved. Published by Tangerine Press, an imprint of Scholastic Inc., *Publishers since 1920*. SCHOLASTIC, TANGERINE PRESS, and associated logos are trademarks and/or registered trademarks of Scholastic Inc.

The publisher does not have any control over and does not assume any responsibility for author or third-party websites or their content.

No part of this publication may be reproduced, stored in a retrieval system, or transmitted in any form or by any means, electronic, mechanical, photocopying, recording, or otherwise, without written permission from the publisher. For information regarding permission, write to Scholastic Inc., Attention: Permissions Department, 557 Broadway, New York, NY 10012.

Copyright © 2019 Scholastic Inc.

10 9 8 7 6 5 4 3 2 1

ISBN: 978-1-338-59875-9

Printed in Ningbo, China

AMAZING SPACE

We live in a place called the universe. What's in the universe? EVERYTHING! You, your pets, and your house are in the universe. Our planet, called Earth, is in the universe. Beyond our planet, there are other planets—along with stars, moons, and much more. Here's the simple rule: If something exists—living or nonliving— it's part of the universe.

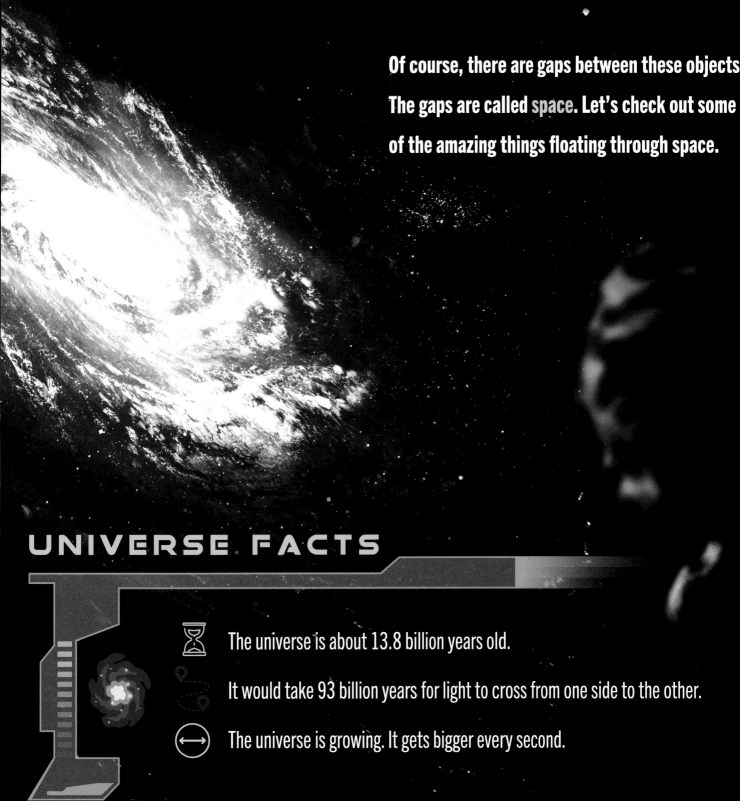

Of course, there are gaps between these objects. The gaps are called space. Let's check out some of the amazing things floating through space.

UNIVERSE FACTS

- The universe is about 13.8 billion years old.
- It would take 93 billion years for light to cross from one side to the other.
- The universe is growing. It gets bigger every second.

OUR SOLAR SYSTEM

MERCURY
VENUS
EARTH
MARS
JUPITER

Our small part of the universe is called the solar system. It's like our neighborhood. A hot star—our sun—sits right in the middle of the solar system. Planets go around and around the sun in circles called orbits. Lots of other things, including comets and asteroids, orbit the sun too. It's a busy neighborhood!

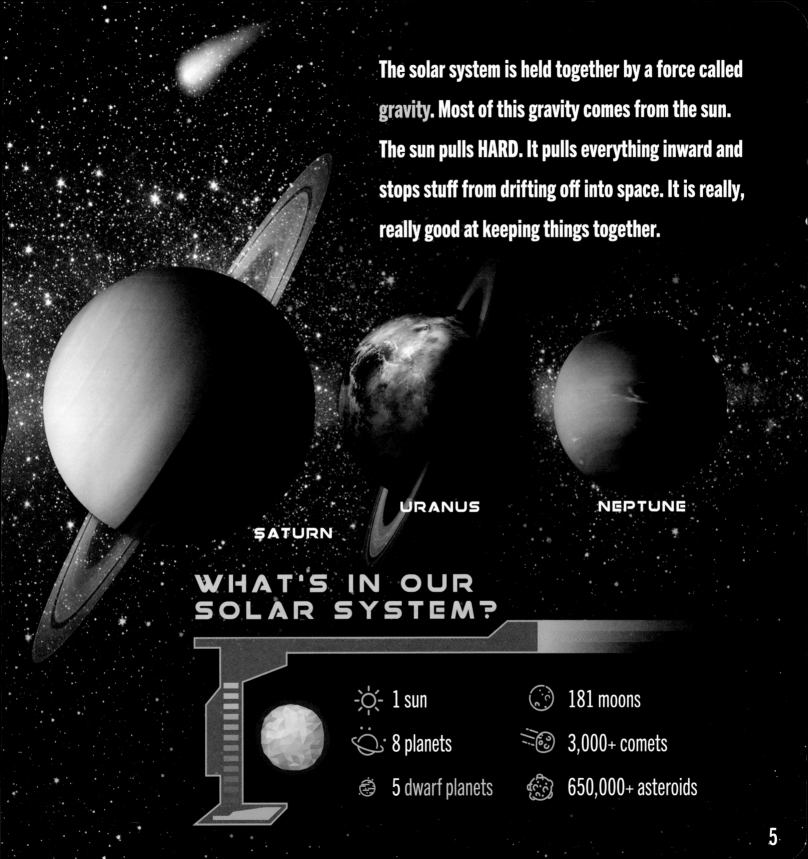

The solar system is held together by a force called gravity. Most of this gravity comes from the sun. The sun pulls HARD. It pulls everything inward and stops stuff from drifting off into space. It is really, really good at keeping things together.

SATURN   URANUS   NEPTUNE

WHAT'S IN OUR SOLAR SYSTEM?

- ☀ 1 sun
- 🪐 8 planets
- 🌑 5 dwarf planets
- 🌖 181 moons
- ☄ 3,000+ comets
- 🪨 650,000+ asteroids

SUN

The sun is the only star in our solar system. It is a big, hot gas ball that blasts heat and light in all directions. Without this heat and light, our solar system would be cold and dark.

The sun is humongous. It is about 864,938 miles (1,392,000 km) across. That is much bigger than our planet. You could fit more than 1 million Earths into the sun.

The sun is also very hot. It is a blistering 10,000 degrees Fahrenheit (5,540 degrees C) on its surface. Inside, things are even hotter. The sun's center reaches about 27 million degrees Fahrenheit (15 million degrees C). That's hot!

SUN FACTS

 Anything solid is called **matter**. The sun contains about 99.8 percent of the matter in our solar system.

 There are cooler spots on the sun's surface. These are called **sunspots**.

 Sometimes gas blasts off the sun's surface. This is called a **solar flare**.

THE INNER PLANETS

Planets are large, round bodies of rock that orbit a sun. Our solar system has eight planets. Some of them are very far away from the sun. Others are closer.

The four planets closest to the sun are called the inner planets. Earth is in this group. The other three planets in the group are like our next-door neighbors! All four inner planets are small and rocky. Let's take a peek at our planetary pals.

MERCURY

Mercury is the closest planet to the sun, and it's FAST. It zips around the sun in just 88 days. Mercury is hot stuff—its surface can be up to 800 degrees Fahrenheit (427 degrees C)!

VENUS

Venus is the second planet from the sun. It is about the same size as Earth. It is covered by a thick blanket of clouds. Warm and cozy!

EARTH

Earth is the third planet from the sun. The surface is a swirl of blue, green, brown, and white. These colors come from water, plants, soil and sand, ice, and clouds.

MARS

Mars is the fourth planet from the sun. It's rusty—really! It is covered with red-brown dirt that has the same chemicals as rust and blood. Mars could use a little bath.

EARTH

Earth is our home planet—our home sweet home! Earth is pretty awesome. It is the only planet in our solar system with open bodies of water. It is also the only planet with oxygen-rich air, and known to have life. That makes Earth very special in our solar system and maybe even in the whole universe. Scientists have looked and looked, but they have not found life anywhere else…yet.

EARTH FACTS

- Earth is sometimes called a Goldilocks planet because it is "just right" for life.
- Earth is more than 4.5 billion years old.
- Earth moves through space at 67,000 miles per hour (107,800 kph).

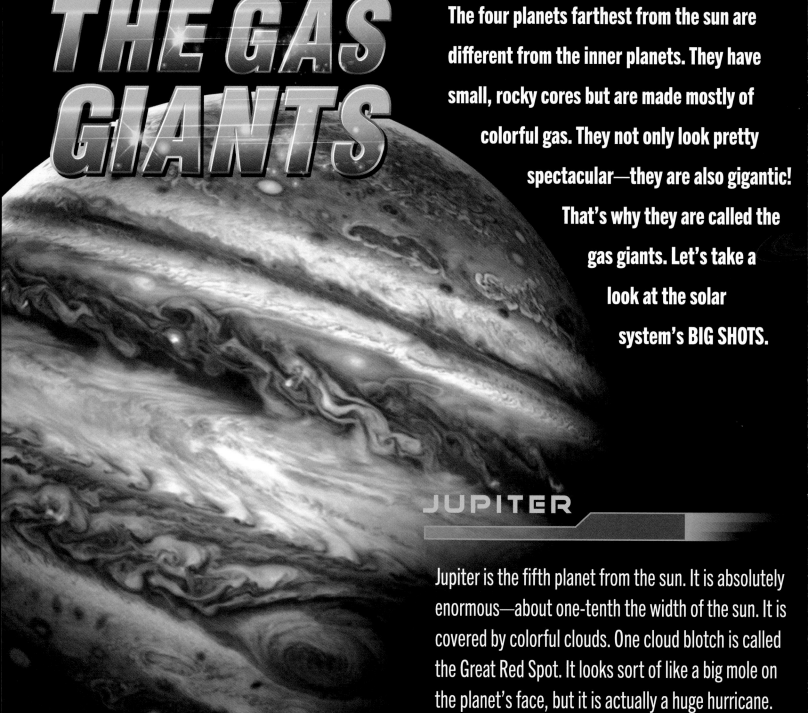

THE GAS GIANTS

The four planets farthest from the sun are different from the inner planets. They have small, rocky cores but are made mostly of colorful gas. They not only look pretty spectacular—they are also gigantic! That's why they are called the gas giants. Let's take a look at the solar system's BIG SHOTS.

JUPITER

Jupiter is the fifth planet from the sun. It is absolutely enormous—about one-tenth the width of the sun. It is covered by colorful clouds. One cloud blotch is called the Great Red Spot. It looks sort of like a big mole on the planet's face, but it is actually a huge hurricane. This storm is as wide as three Earths lined up in a row!

SATURN

Saturn is the sixth planet from the sun. You can't miss it—it's the one wearing a ring. Lots of them, actually! The planet is circled by massive bands of rock and dust. The rings are thousands of miles wide but only about 66 feet (20 m) thick. That's about the length of five cars lined up in a row. It's not very thick considering how huge the rings look!

URANUS

Uranus is the seventh planet from the sun. It is the coldest planet in our solar system. Uranus is sometimes called the "Ice Giant" because it has a layer of ice around its chilly core. It has thirteen rings and spins on its side.

NEPTUNE

Neptune is the eighth planet from the sun. It is deep blue all over, like the deep blue sea. But it's not water. The color comes from a gas that reflects blue light. Neptune is a very windy place. It has great storms with wind speeds more than 1,100 miles per hour (1,770 kph).

DWARF PLANETS

Besides the eight main planets, our solar system has five dwarf planets that orbit the sun. Dwarf planets are too small to be regular planets. They are too big to be space rocks. They are right in the middle! The dwarf planets are named Ceres, Haumea, Makemake, Eris, and Pluto.

DWARF PLANET FACTS

 Pluto and Eris are the biggest dwarf planets. They are almost the same size.

 Ceres is the smallest dwarf planet.

 There are probably many more dwarf planets in our solar system. Scientists are still looking.

Pluto was once considered our solar system's ninth planet. But scientists changed this after they discovered other dwarf planets. They thought Pluto was more like a dwarf planet than a regular planet, so they changed its title.

CERES

MAKEMAKE

HAUMEA

ERIS

PLUTO

MOONS

Moons are bodies that orbit planets. The smallest moon in our solar system is named Deimos. It orbits Mars and it is just 7 miles (11.3 km) across. The biggest moon is named Ganymede. It orbits Jupiter. It is bigger than the planet Mercury!

Some planets have many moons. Others have none. Earth has just one moon, but it's a pretty big one! It is the fifth biggest moon in our solar system.

HOW MANY MOONS?

Planet	Moons	Planet	Moons
Mercury	0	Jupiter	79
Venus	0	Saturn	62
Earth	1	Uranus	27
Mars	2	Neptune	14

DISTANT STARS

Outside our solar system, there are other stars. They gather in clusters called galaxies. There are trillions of galaxies in the universe! Each galaxy can contain billions of stars. That's a whole lot of star power!

Like our sun, all of these stars are HOT STUFF. They give off heat and light. We cannot feel the heat from distant stars—they are too far away. But we can see their light. We can also look through telescopes to spy on faraway stars and galaxies.

STAR FACTS

Our home galaxy is called the Milky Way. It contains at least 100 billion stars.

Other than the sun, Earth's nearest star is Proxima Centauri.

Many stars outside our solar system have their own planets. These are called exoplanets.

DIFFERENT STARS

Stars are like people: every one is different! For one thing, stars come in many sizes. Some stars are much, much bigger than our sun. Others are smaller. Compared to most stars, our sun is on the small side.

Stars also come in many different colors. The biggest, hottest stars shine blue. Slightly cooler stars are white. Even cooler stars are yellow. The coolest stars of all are deep orange to red in color. The universe is a very colorful place.

BLUE GIANTS ARE HUGE, HOT STARS.

LOOKING INTO THE PAST

Stars make light, and light is an amazing thing. It streams through space at mind-bending speeds. Light travels 186,000 miles (300,000 km) in one second. That's like seven and a half times around Earth! In one second! Imagine that!

Even at this speed, it takes 8 minutes and 20 seconds for light to travel from the sun to Earth. It can take billions of years from the farthest stars. We see this light today, but it was made long, long ago. So seeing it is like looking into the past.

LIGHT FACTS

 Light travels about 6 trillion miles (9.7 trillion km) in one year.

 This distance is called a light year.

 The farthest individual star we can see is 9 billion light years away.

 The farthest galaxy we can see is 13.3 billion light years away.

Our home galaxy is 100,000 light years across.

 Scientists think the entire universe is about 93 billion light years across!

STARS' LIFE CYCLE

Stars have a life cycle. They are born, they live for a while, and then they die. Stars form in clouds of gas and dust. A cloud like this is called a nebula. Sometimes gravity pulls a nebula's gas and dust into a clump. The clump gets hot and starts to burn. A star is born!

Stars burn for billions of years. At some point they burn up all their gas. When this happens, big changes start to occur. Some stars shrink and get dimmer. Some go dark. Some explode! Some collapse inward and get very dense and heavy.

EXPLODING STARS

A star explosion is called a supernova.

Huge stars can become black holes after they explode. A black hole's gravity is very strong. It sucks in everything, even light.

Other stars can become neutron stars after they explode. Neutron stars are small and very dense.

WHAT ARE CONSTELLATIONS?

Twinkle, twinkle, little star! Look up at the sky on a clear night. You will see points of light everywhere. Most of them are stars. A few are distant galaxies and nebulae. Using just our eyes, we can see about 9,000 points of light from Earth.

These dots of light seem to form connect-the-dot patterns. These patterns are called **constellations**. There are 88 official constellations. They form the shapes of animals, people, and objects. Every constellation has a name.

CONSTELLATION FACTS

Forty-eight constellations were first listed in the second century by an astronomer named Ptolemy.

Forty new constellations were adopted in 1922. Scientists wanted to include every star in the sky!

The biggest constellation is called Hydra. It is a water snake.

The smallest constellation is called Crux. It is a cross.

27

WELL-KNOWN CONSTELLATIONS

Here are some well-known constellations. Have you seen them?

Look for these patterns in the night sky.

URSA MAJOR AND URSA MINOR

Ursa Major and Ursa Minor are bears. The seven stars in each bear look like ladles. They are called the Big Dipper and the Little Dipper.

ORION AND TAURUS

Orion is a hunter. This constellation looks like a man holding a sword and shield. The three stars in the middle are his belt. Orion is hunting the constellation Taurus, which is a bull.

SKYWATCHING

Earth's skies are full of wonders. Look up on a clear night and check out our incredible space neighborhood. What do you see in the night sky?

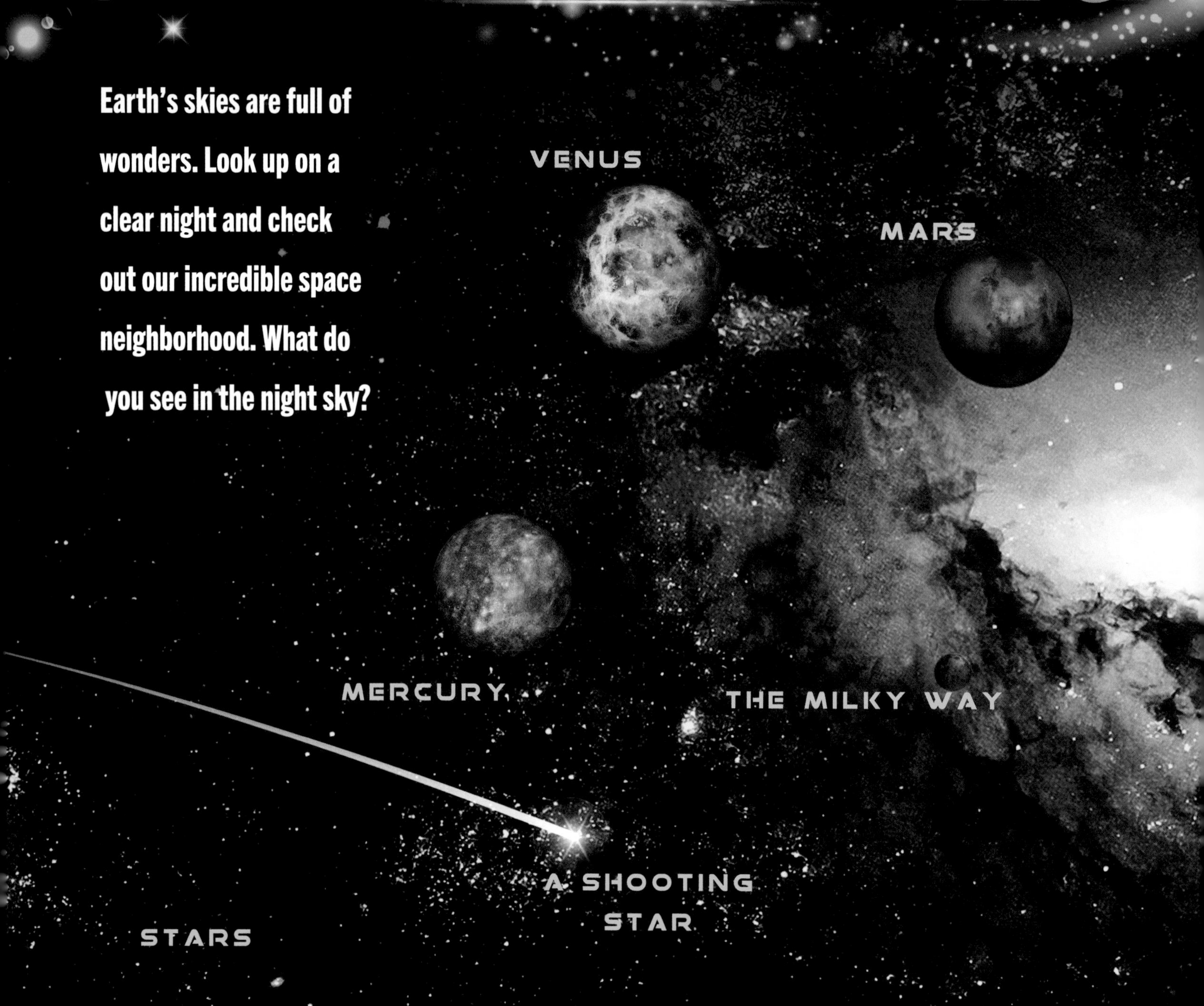

VENUS

MARS

MERCURY

THE MILKY WAY

A SHOOTING STAR

STARS

GLOSSARY

Black holes: collapsed stars with very strong gravity

Constellations: star patterns that seem to make pictures

Dense: tightly packed

Dwarf planets: small planet-like bodies

Exoplanets: planets that orbit stars outside our solar system

Galaxies: clusters of billions of stars

Goldilocks planet: planet that is "just right" for people to survive

Gravity: a force that pulls on things

Light year: the distance light travels in one year

Matter: anything solid

Milky Way: our home galaxy

Moon: a body that orbits a planet

Nebula: a huge cloud of dust and gas in space

Neutron stars: small, dense bodies that remain after a large star explodes

Orbit: to travel around and around an object

Oxygen: a gas that people must breathe to survive

Planets: large, rounded bodies that orbit a star

Solar flare: a blast of gas from the sun's surface

Space: the gaps between objects in the universe

Star: a huge ball of bright, hot gas

Sunspots: cool spots on the sun's surface

Supernova: an explosion when a star reaches the end of its life

Telescopes: instruments that make distant objects look bigger

Universe: everything that exists

Yellow dwarf: a type of star; our sun is a yellow dwarf

METRIC TABLE

The metric system is a system of measurements. It is used in many parts of the world. It is also used by all scientists, no matter where they live. Here are some common abbreviations for metric measurements.

C = Celsius	**kph** = kilometers per hour
km = kilometers	**m** = meters